HOW TO MASTER SECRET WORK

The Communist Revolutionary's Underground
Operations Manual from Apartheid-era South
Africa

Originally published in the South African
Communist Party's Underground Newspaper
Umsebenzi between 1985 and 1987

CONTENTS

1. Introduction 5
2. Setting Up a Secret Network 8
3. Some Rules of Secrecy 11
4. Surveillance 14
5. Surveillance Techniques 17
6. Surveillance Techniques (Continued) 20
7. Counter-Surveillance 23
8. The Check Route 26
9. Check Route With Assistance and By Vehicle 30
10. Checking by Car 33
11. Cutting the Tail 36
12. Secret Communications 39
13. Personal Meetings 42
14. Emergency and Check Meetings 45
15. Blind Meeting 49
16. Non-Personal Communication 52
17. Signals 55
18. Dead Letter Box 58
19. Stationary, Portable and Mobile DLBs 63
20. Failure and How to Deal With It 66
21. Detecting and Localising Failure 69

HOW TO MASTER SECRET WORK

1. INTRODUCTION

This is a pamphlet about the role of secrecy in solving the tasks of the Revolution. Secrecy gives us protection by starving the enemy of information about us. Secrecy helps us build a strong revolutionary movement to overthrow the enemy.

There is nothing sinister about using secret methods to help win freedom. Through the ages the ruling classes have made it as difficult as possible for the oppressed people to gain freedom. The oppressors use the most cruel and sinister methods to stay in power. They use unjust laws to ban, banish, imprison and execute their opponents. They use secret police, soldiers, spies and informers against the people's movements. But the people know how to fight back and how to use secret methods of work.

The early history of struggle in our country is full of good examples. Makanda, Cetshwayo, Sekhukhune and Bambatha made use of secret methods to organise resistance. Bambatha, for example, prepared his rebellion against colonialism in great secrecy from the Nkandla forest.

Secrecy has Helped us Outwit the Enemy:

The enemy tries to give the impression that it is impossible to carry out illegal work. The rulers boast about all our people they have killed or captured. They point to the freedom fighters locked up in the prisons. But a lot of that talk is sheer bluff. Of course it is impossible to wage a struggle without losses. The very fact, however, that the South African Communist Party and African National Congress have survived years of illegality is proof that the regime cannot stop our noble work. It is because we have been mastering secret work that we have been able, more and more, to outwit the enemy.

Discipline, Vigilance and Self-Control:

Secret methods are based on common sense and experience. But they must be mastered like an art. Discipline, vigilance and self-control are required. A resistance organiser in Nazi-occupied France who was never captured said this was because he 'never used the telephone and never went to public places like bars, restaurants and post offices'. He was living a totally underground life. But even those members of a secret movement who have a legal existence must display the qualities we have referred to.

Study and Apply the Rules of Secrecy:

Most people know from films and books that secret work involves the use of codes, passwords, safe houses and hiding places. Activists must study the rules of secrecy and apply them seriously. This enables us to build up secret organisations linked to the people. This secret network becomes a vital force in helping to lead the people in the struggle for power. In our series we will discuss such topics as:

1. How to set up a secret network;

2. The rules of secrecy;

3. How to overcome surveillance (i.e. observation);

4. Secret forms of communication;

5. Technical Methods such as secret writing, hiding places etc.;

6. How to behave under interrogation (i.e. when being questioned by the enemy). These are among the main elements of secret work.

To organise in secret is not easy, but remember: The most difficult work is the most noble!

2. SETTING UP A SECRET NETWORK

We have said that secret work helps us overcome the problems created by the enemy. This helps in the vital task of building an underground organisation or secret network. The network must lead the people in the struggle for power. It does not compete with the progressive legal organisations but reinforces them. Let us look at some of the main measures involved:

1. Only serious and reliable people can be included in the secret network. The leaders must study the potential recruits very carefully. They are looking for people who are politically clean, determined, disciplined, honest and sober. People who can keep a secret. People who are brave and capable of defying the enemy even if captured.

2. Recruits are organised into a unit or cell of three or four people. The number is limited in case of failure or arrest. The cell leader is the most experienced person. The cell members must not know the other members of the network.

3. Only the cell leader knows and is in contact with a more senior member of the network. This senior contact gives instructions from the leadership and receives reports.

4. A small committee of the most experienced people leads the network. This is a leadership cell of two or three

persons. This cell might be in charge of a factory, location, township or city. A city network takes the form of a pyramid. The city underground committee is at the top. Local cells are at the base. Middle command cells are in between. Start with one cell. Gain experience before building more.

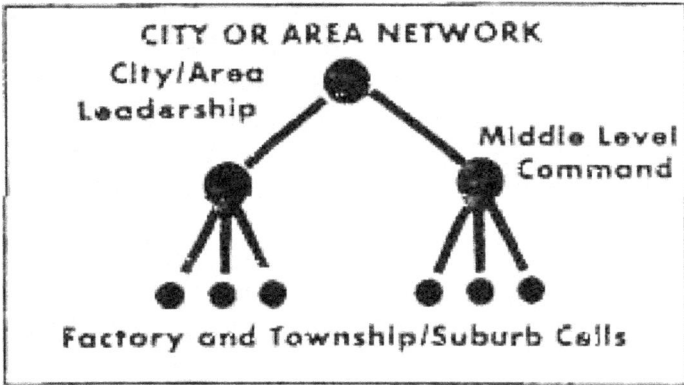

CITY OR AREA NETWORK
City/Area Leadership
Middle Level Command
Factory and Township/Suburb Cells

5. A rule of secret work is that members must know only that which is necessary to fulfil their tasks. Everyone, from top to bottom, must have good cover stories to protect them. This is a legend or story which hides or camouflages the real work being done. For example: a secret meeting in a park is made to look like a chance meeting between friends. If they are ever questioned they give the legend that they simply bumped into each other and had a discussion about football.

6. All members of the network are given code names. These conceal their real identities. They must have good identification documents. Especially those living an illegal life. A lot of time and effort must be given to creating good legends to protect our people. There is nothing that arouses suspicion as much as a stranger who has no good reason for being around.

7. All illegal documents, literature, reports and weapons (when not in use) must be carefully hidden. Special hiding places must be built. Codes must be used in reports to

conceal sensitive names and information.

8. The leaders must see that all members are trained in the rules and methods of secret work . It is only through this training that they will develop the skills to outwit the enemy.

9. Technical methods such as the use of invisible writing, codes and disguise must be mastered. Counter-surveillance methods which help check whether one is being watched by the enemy must be known. Secret forms of communicating between our people must be studied and used. This is all part of the training. These methods will be dealt with later.

10. Specialisation: Once the network has been developed some cells should specialise in different tasks such as propaganda, sabotage, combat work, mass work, factory organisation etc.

In the meantime you can start putting into practice some of the points already dealt with. Begin to work out legends in your work. What innocent reason can you give if a friend or a policeman finds this journal in your possession?

3. SOME RULES OF SECRECY.

Carelessness leads to arrests. Loose talk and strange behaviour attracts the attention of police and izimpimpi. Secret work needs vigilance and care. Rules of secrecy help to mask our actions and overcome difficulties created by the enemy. But first let us study the following situation:

What Not To Do

X, a trade unionist, also leads a secret cell. He phones Y and Z, his cell members, and arranges to meet outside a cinema. X leaves his office and rushes to the meeting 30 minutes late. Y and Z have been anxiously checking the time and pacing up and down. The three decide to go to a nearby tea-room where they have often met before. They talk over tea in low tones. People from the cinema start coming in. One is a relative of X who greets him. Y and Z are nervous and abruptly leave. When X is asked who they were he hesitates and, wanting to impress his relatives, replies: *'They're good guys who like to hear from me what's going on'*. This opens the way for a long discussion on politics. X has made many errors which would soon put the police on the trail of all three. These seem obvious but in practice many people behave just like X. They do not prepare properly; rush about attracting attention; fail to keep time; do not cover the activity with a legend (cover story); talk loosely etc. Others pick up the bad style of work. X should set a good example for Y and Z.

To avoid such mistakes rules of secrecy must be studied and practised. They might seem obvious but should never be taken for granted.

Things to Remember

1. Always have a believable' legend to cover your work! (X could have said Y and Z were workers he vaguely knew whom he had met by chance and had been encouraging to join the union).

2. Underground membership must be secret! (X had no need to refer to Y and Z as '*good guys*').

3. Behave naturally and do not draw attention to yourself! 'Be like the people'. Merge with them! (X, Y and Z behaved suspiciously.)

4. No loose talk! Guard secrets with your life! Follow the saying: 'Don't trust anyone and talk as little as possible'. (X fails here).

5. Be vigilant against informers! They try to get close to you, using militant talk to 'test' and trap you. (Can X be so sure of his relative?)

6. Be disciplined, efficient, punctual (X was none of these). Only wait ten minutes at a meeting place. The late comer may have been arrested.

7. Make all preparations beforehand! Avoid a regular pattern of behaviour which makes it easy for the enemy to check on you. (X made poor arrangements for the meeting; rushed there from a sensitive place and could have been followed; used the tea-room too often).

8. Do not try to discover what does not concern you! Know only what you have to know for carrying out your tasks.

9. Be careful what you say on the phone (which may be 'bugged'), or in a public place (where you can be overheard)! Conceal sensitive information such as names etc. by using simple codes!

10. Remove all traces of illegal work that can lead to you! Wipe fingerprints off objects. Know that typewriters can be traced; goods bought from shops can be checked.

11. Hide materials such as leaflets, weapons etc! But not where you live. Memorise sensitive names, addresses etc. Don't write them down!

12. Carry reliable documents of identification!

13. Know your town, its streets, parks, shops etc. like the palm of your hand! This will help you find secret places and enable you to check whether you are being followed.

14. If you are arrested you must deny all secret work and never reveal the names of your comrades even to the point of death!

15. Finally, if any member of your underground cell is arrested, you must immediately act on the assumption that they will be forced to give information. This means taking precautions, such as going into hiding if necessary.

When the rules of secrecy are practised revolutionaries make good progress. Practice makes perfect and with discipline and vigilance we will outwit the enemy and we will win!

4. SURVEILLANCE

What is Surveillance?

In their efforts to uncover secret revolutionary activity the police put a close watch on suspected persons and places. This organised form of observation is called surveillance. There are two general types of surveillance: mobile and stationary. Mobile is sometimes refer red to as 'tailing' or 'shadowing' and involves following the suspect (subject) around. Stationary is observing the subject, his or her home and workplace, from a fixed position. This can be from a parked car, neighbouring building or shop and is referred to as a 'stake-out' in detective films. Surveillance combines both 'tailing' and 'stake-outs'.

Counter-Surveillance

Members of a secret network must use methods of counter-surveillance to protect themselves and their underground organisation. You can establish whether you are being watched or followed. These methods can be effectively used and help you to give the police the impression that you are not involved in secret work. Before considering these methods of protection, however, we need to be more aware of the enemy's surveillance methods. For it is not possible to deal with surveillance unless we know how it operates.

Aim of Surveillance

The primary aim of surveillance is to gather information about the subject and to check out whether he or she is involved in secret work. The police seek to establish the links between the subject and those he or she might be working with. The enemy wants to identify you and locate the residences and secret places you use. They try to collect evidence to prove that illegal work has been committed. An important use of surveillance is to check on information received from informers.

Decision for Surveillance

A decision to place a subject under surveillance is taken at a high level. The decision will include the intensity and duration for example whether for 8, 16 or 24 hours per day over a period of one, two, three or more weeks. The decision will involve placing the subject's house and workplace under observation and having his or her phone tapped either temporarily or permanently. The number of persons involved in the operation will be decided upon and they will be given the known facts about the subject including a description or photograph. Whether the surveillance ends with the arrest of the subject will depend on what is learnt during the investigation.

The Surveillance Team

Specially trained plainclothes men and women are used to carry out surveillance. Their identities are kept strictly secret. They are not the normally known or public special branch policemen. They are aged between 25 and 50 years and have to be physically fit for work. In appearance and dress they are average types. They try to blend in with their surroundings and avoid drawing attention to themselves. For example, smartly dressed whites will not be used to follow a black person in a poor, run-down area.

A team may consist of 2-4 people with a car in support. Usually one team is used at a time but more will be deployed if required. The subject will be followed by foot, car or public transport if necessary. The surveillants communicate with each other by discreet hand signals and small radio transmitters. They make minor changes in their clothing and appearance to help prevent recognition. For the same reason they try to avoid abrupt and unnatural movements when following the subject.

In a crowded city street they will 'stick' close to the subject (within 20 metres) for fear of losing him or her. In a quiet residential area they will 'hang' back (over 50 metres) for fear of exposing themselves. They have set plans and procedures for 'tailing' the subject which involves the constant interchanging of positions. It is important to know these various techniques of foot and vehicle surveillance.

5. SURVEILLANCE TECHNIQUES

We have defined surveillance as an organised form of observation in which the police put a close watch on suspected persons or places. Various types of surveillance and techniques of 'tailing' the suspect (subject) are used. A subject's home or place of work might be under observation from a stationary or 'fixed' position such as a neighbouring residence or vehicle. All comings and goings are recorded. When the subject leaves his or her home they may be followed by foot or car or combination of both. All the places they visit and people they meet are noted, photographed and followed too if necessary.

Foot Surveillance

At least two people will be used to follow the subject whom we will call 'S'. They will communicate through hand-signals and 'walkie-talkie' radios so as to guide and assist each other. They will keep as close to S as 15 metres in crowded areas and hang well back, up to 100 metres, in quiet streets. They will try to be as inconspicuous as possible so as not to arouse S's suspicions. They will have a car to assist them, which keeps out of sight in the adjacent streets.

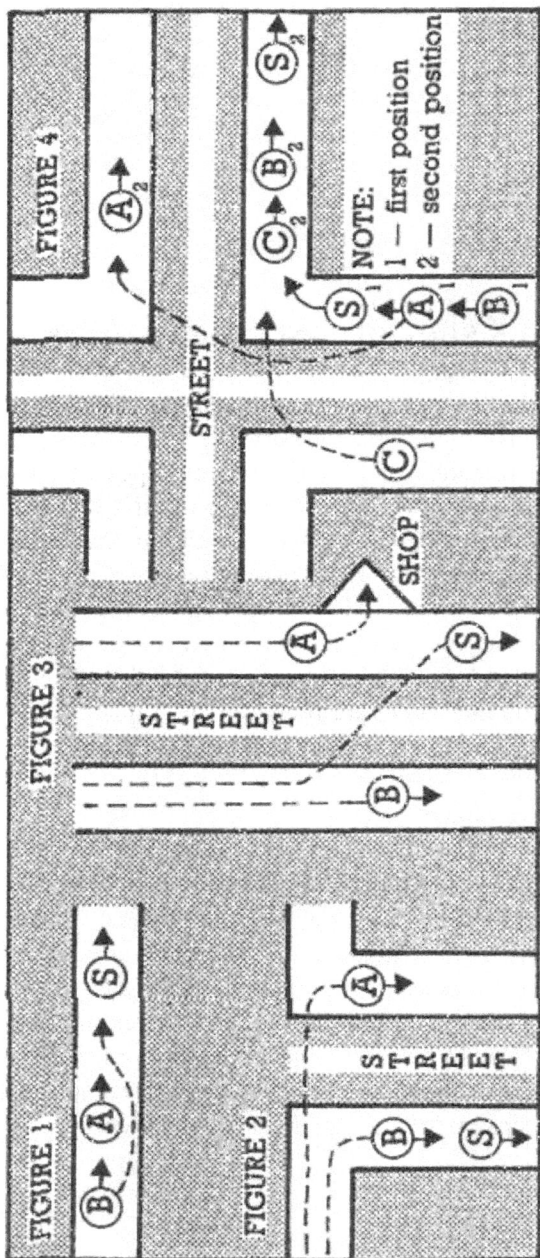

FIGURE 4

FIGURE 3

FIGURE 1

FIGURE 2

STREET

STREET

STREET

SHOP

NOTE:
1 — first position
2 — second position

Two-Man or 'AB' Surveillance

The person following directly behind S is A. The second person is B, who follows on behind A, as if in a chain. A and B alternate positions, 'leap-frogging' over each other (Figure 1). When S turns right at a corner A drops back out of sight and B takes the lead position. An alternative technique is for A to cross the road and then turn right. In this case A is not now following directly behind B as in a chain, but is parallel to B on the opposite side of the road to both B and S and slightly to their rear (Figure 2). A and B will avoid direct contact with S. If S now crosses the street to the left A will either fall back, enter a shop or walk swiftly ahead, while B will follow S from his side of the street (Figure 3).

Three Man or 'ABC' Surveillance

Inclusion of the extra man makes tailing S easier. A follows S, B follows A and C operates across the street from S to the rear. When S turns a corner, A may continue in the original direction, crossing the street instead of immediately turning. A thus takes the C position, whilst either B or C can take A's original position (Figure 4).

A variety of techniques can obviously be used. But the idea is generally the same. Those following must keep the subject under constant observation without arousing suspicion. The more persons used, the greater the scope and flexibility of the operation.

Remember: By knowing the methods of the enemy we can deal with him and defeat him! (Diagram 2)

We have dealt above with following people on foot. We now turn to 'tailing' by vehicle.

6. SURVEILLANCE TECHNIQUES (Continued)

Vehicle Surveillance

A variety of vehicles may be used in surveillance car, van, truck or motorbike. These must be dependable and powerful but not flashy so as to avoid attracting attention. A surveillance vehicle will carry no visible police identification but of necessity will be equipped with a two-way radio (so look out for the antenna!)

In heavy traffic the tailing vehicle will stick close behind the suspect's vehicle, hereafter referred to as the subject or 'S'. In light traffic it will hang well back, but it will always try to keep two or three cars behind S (Figure 1), especially in One-Vehicle Surveillance. The tailing-vehicle will remain in the same lane as S to avoid making sudden turns from the wrong lane. There are normally two persons in a tailing vehicle. The passenger is always ready to alight and carry out foot surveillance if S parks his or her car or gets out of it. As in foot surveillance, inconspicuous actions are required so as not to arouse the suspicions of S. When more tailing vehicles are used, the scope and flexibility of the operation is increased. But normally two tailing vehicles are utilised. The number depends on the degree of urgency of the operation.

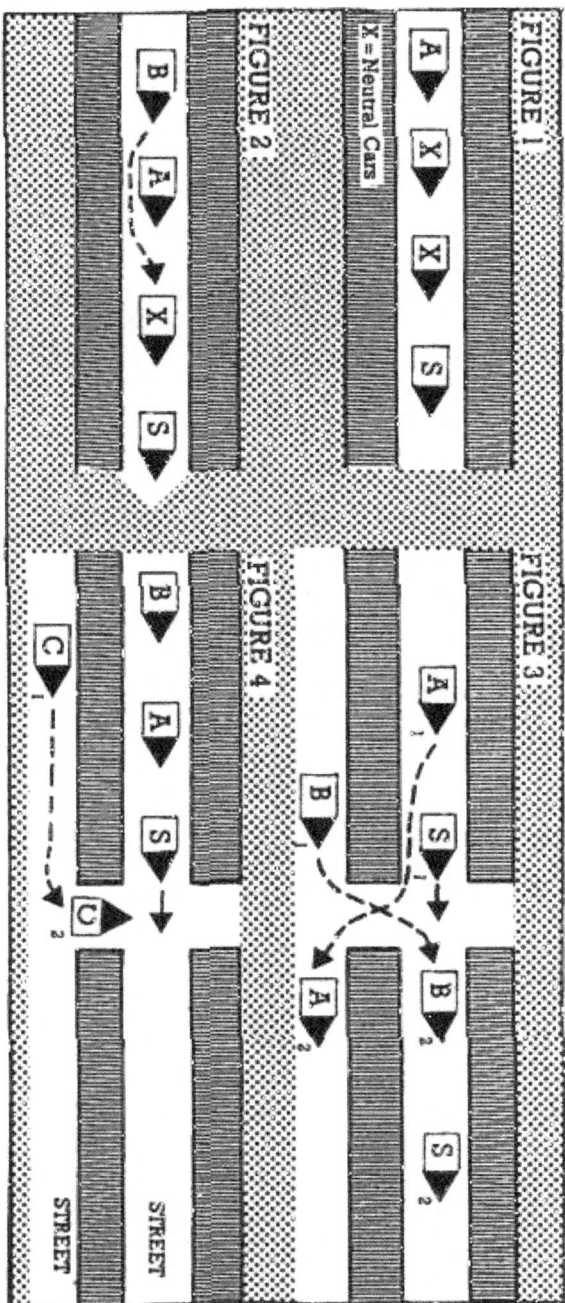

FIGURE 1

X = Neutral Cars

A ▶ X ▼ X ▼ S ▼

FIGURE 2

B ▼ A ▼ X ▼ S ▼

FIGURE 3

A ▼ S ▼ B ▼ B ▼ A ▼ S ▼

FIGURE 4

B ▼ A ▼ S ▼ C ▼ C ▶

STREET STREET

Two and Three Vehicle Surveillance

When two tailing vehicles are used, the lead tail A will remain two or three cars behind S and B will remain behind A, as in a chain. They will always keep switching places (Figure 2). When using a parallel tailing technique, A remains behind S and B keeps pace in a parallel street. A and B keep switching positions (Figure 3). With three tailing vehicles the possibilities are increased. A and B follow S in a chain and a third vehicle C travels in a parallel street. C may even speed ahead of S, awaiting it at an intersection before falling in behind and taking A's position. This allows A to turn off and follow in a parallel street (Figure 4).

Reflectors and Bleepers

Those carrying out surveillance may try to place a strip of reflectorised tape on the rear of the subject's vehicle or break a tail-light to make it easier to spot it at night. Or they may place an electronic tailing device on S's car, called a Bumper Bleeper. This is a small metal box which can be fixed to the vehicle with magnets in seconds. A radio signal is transmitted which the tailing vehicle picks up on a receiver. S's car can be tracked even when out of view! Such gadgets do not, however, make it impossible to avoid being tailed. It only means that you must be alert and check for such devices. Knowing it is there can help you to really mislead the enemy!

Progressive Surveillance

This technique is used when extreme caution is needed because the subject is likely to use all methods to uncover possible surveillance. S is only followed for a limited distance each day by foot or car. Observation is picked up again at the time and place where it was previously discontinued. This continues day after day until surveillance is completed or discontinued. Remember! Know the enemy's methods to deal with him and defeat him!

7. COUNTER-SURVEILLANCE

We have been examining the enemy's surveillance methods, that is, the forms of observation used to watch suspects and uncover secret revolutionary activity. We now turn to counter-surveillance, which is the methods we use to deal with enemy observation.

Qualities Needed

For successful counter-surveillance you need to be aware of your surroundings and be alert to what is going on round you. That means having a thorough knowledge of the town or area in which you live and work and knowing the habits of the people. You need basic common sense, alertness and patience together with cool and natural behaviour and a knowledge of certain tactics or ruses (which will be discussed later). It is important not to draw attention to oneself by strange behaviour such as constantly looking over one's shoulder. And one must guard against paranoia, that is, imagining that everyone you see is following you. It is necessary to develop powers of observation and memory (which come with practice) so that you notice what is usual and remember what you have seen. It is when you notice the same person or unusual behaviour a third or fourth time that you are able to conclude that it adds up to surveillance and not coincidence.

Are You Being Watched?

Study the normal situation where you live, work and socialise so as to immediately recognise anything out of the ordinary. Are strangers loitering about the streets? Are strange cars parked where the occupants have a commanding view of your home? They may be a distance away spying on you through binoculars. Do the vehicles have antennae for two-way radio communication? Do you notice such strangers or vehicles on several occasions and in other parts of the town? This would serve to confirm interest in you.

Have strangers moved into neighbouring houses or flats? Do you notice unusual comings and goings or suspicious movements at upstairs windows? Try discreetly to check who such people are. The enemy might have created an observation post in the house opposite the road or placed an agent in the room next door to you! Be sensitive to any change in attitude to you by neighbours, landlady,shopkeeper etc. The enemy might have mobilised them for surveillance. Know such people well, including the local children, and be on good terms with all. Then if strangers question them about you, they will be more inclined to inform you.

Know the back routes and concealed entrances into your area so that you may slip in and out unnoticed. Secretly check what is going on in the vicinity after pretending to retire for the night. Avoid peering from behind curtains, especially at night from a lit room. This is as suspicious as constantly glancing over one's shoulder and will only alert the enemy to conceal themselves better.

Record all unusual incidents in a note book so you can analyse events and come to a conclusion. Be alert with persons you mix with at work or socially, and those like receptionists, supervisors, waiters and attendants who are well-placed to notice one's movements.

Telephone and Mail

Phone tapping often causes faults. Check with neighbours whether they are having similar problems or is your phone the exception. Is your post being interfered with? Check dates of posting, stamp cancellation and delivery and compare the time taken for delivery with your friends. Examine the envelopes to check whether they have been opened and glued down in a clumsy way. Some of these checks do not necessarily confirm that you are being watched but they alert you to the possibility. To confirm whether you are in fact under observation requires techniques of checking which we will examine next.

8. THE CHECK ROUTE

The Check Route is a planned journey, preferably on foot, along which a person carries out a number of discreet checks in order to determine whether they are under surveillance. These checks take place at predetermined check points which must give you the opportunity of checking for possible surveillance without arousing the suspicion of those tailing you.

The check route should cover a distance of 3-4km, include such activities as shopping, making innocent enquiries, catching a bus, enjoying refreshments etc, and should last about one hour. The route should include quiet and busy areas bearing in mind that it is easier that you have a valid reason for your movements. If your actions are strange and inexplicable you will arouse the suspicions of those following you.

Here is an example of a typical check route. Shortage of space obliges us to confine the check points into a smaller area just a few city blocks than would actually be the case. Check points are numbered 1 to 12.

1. X walks down the street and pauses at a cinema to examine the posters -this gives a good chance to look back down the street and to notice those passing by (without looking over his shoulder),

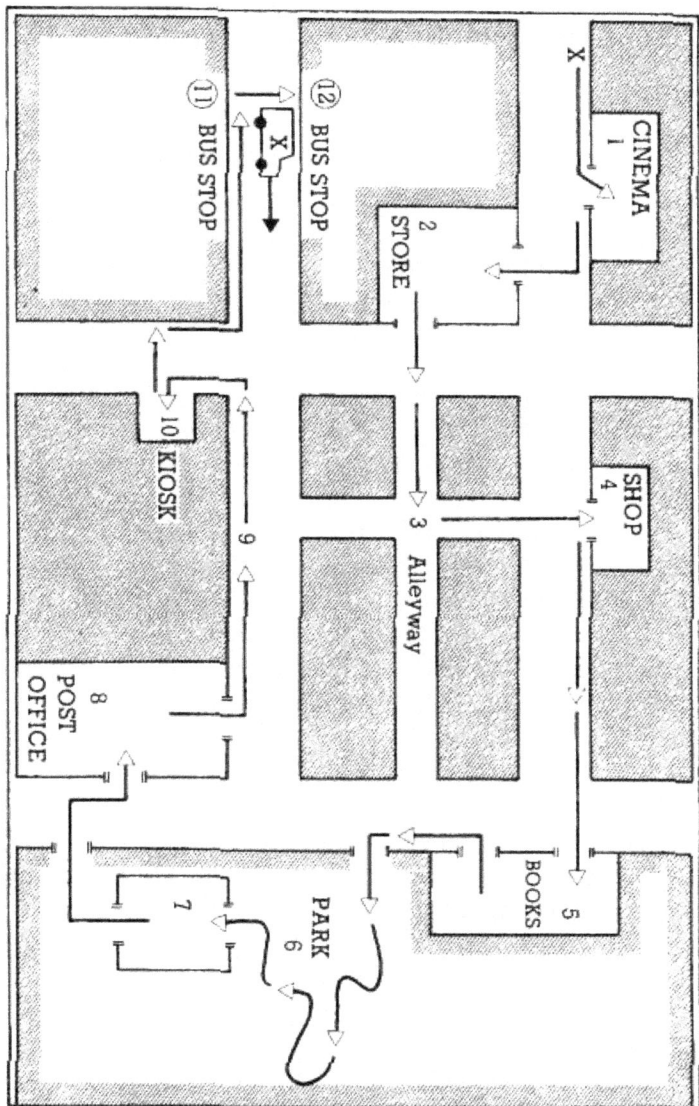

2. X crosses the road looking right and left and pops into a large store; he positions himself near the entrance whilst appearing to examine goods on display; he notices anyone entering after him; wanders around the store using lift, stairways etc. in order to spot anyone paying special interest in him; departs at side exit

3. and crosses street into little-used alleyway or arcade; here he slightly picks up speed and crosses street, where

4. shop with large plate glass windows gives good reflection of alley out of which he has emerged; X notices whether anyone is coming out of that alley to catch up with him ...

5. X now proceeds down the street into bookshop with commanding view of the street he has come down; he browses around noticing anyone entering after him; he also observes whether anyone examines the books he has been browsing through (for a tail would want to check whether X has left a secret communication behind him for a contact); X makes a small purchase and exits...

6. enters park and walks along winding paths which give good view of rear; X throws away an empty cigarette pack and retires to ...

7. an out-door restaurant where he takes his tea; he observes whether anyone picks up the cigarette pack which a tail would want to check as in 5; and notices the customers arriving after him; any tail would want to check whether X is meeting someone; as X leaves he notices whether any of the customers are eager to leave immediately after him ...

8. X crosses the street into a Post Office; once inside he is able to observe whether anyone is crossing the street from the park after him; he buys some stamps and notices anyone queuing behind him (a tail will be especially interested in transactions taking place in post offices, banks etc.); X may also make a

'phone call at a public box and check whether anyone attempts to overhear his conversation;

9. on departing X stops a stranger in the street to ask him the way; this allows him to check whether anyone has followed him out of the Post Office; a tail would also show interest in this stranger (who might be X's contact) and a member of the surveillance team might follow this stranger';

10. X continues down the street, turns sharply at the corner and abruptly stops at a cigarette kiosk; anyone following will most likely come quickly around the corner and could become startled on finding X right in his path.

11. -12. X crosses the street and joins the queue at a bus stop (11) noticing those joining the queue after him; a bit of acting here gives the impression that X is unsure of the bus he wants to catch; he could allow a couple of buses to go by noticing anyone who is doing the same; as a bus arrives at the stop across the road (12), X suddenly appears to realise it is his and dashes across the road to catch it as it pulls away;
 X is alert to anyone jumping on the bus after him and will also pay attention to whoever gets on at the next few stops.

Such a series of checks must be carried out immediately prior to any sensitive appointment or secret meeting. If nothing suspicious has occurred during the Check Route X proceeds to his secret appointment or mission. If, on the other hand, X has encountered certain persons over and over again on the Check Route he will assume he is under surveillance and break his appointment. Bear in mind that anyone following you, even professionals, may become indecisive or startled should your paths unexpectedly cross. A Check Route should also be carried out from time to time to check whether a person is 'clean' or not.

9. CHECK ROUTE WITH ASSISTANCE AND BY VEHICLE

Check Route is a planned journey, the object of which is to check whether you are being followed. The previous example was a check route on foot, by a person acting alone.

With assistance from comrades the exercise becomes more effective. The exercise follows similar lines as previously outlined except that a comrade is stationed at each check point and observes whether anyone is following you as you pass by. It is essential that your behaviour appears normal and does not look as though 'checking' is taking place.

Let us suppose that you are X. Comrades Y and Z position themselves at check points Y1 and Z1 respectively. These observation points must give a good view of your movements, but keep the comrades hidden from enemy agents who might be tailing you. After X passes each check point the comrades move to new positions, in this case Y2 and Z2. They may in fact cover four to five positions each and the whole operation should take one to two hours over an area of three or four kilometres. Comrades must take up each position in good time.

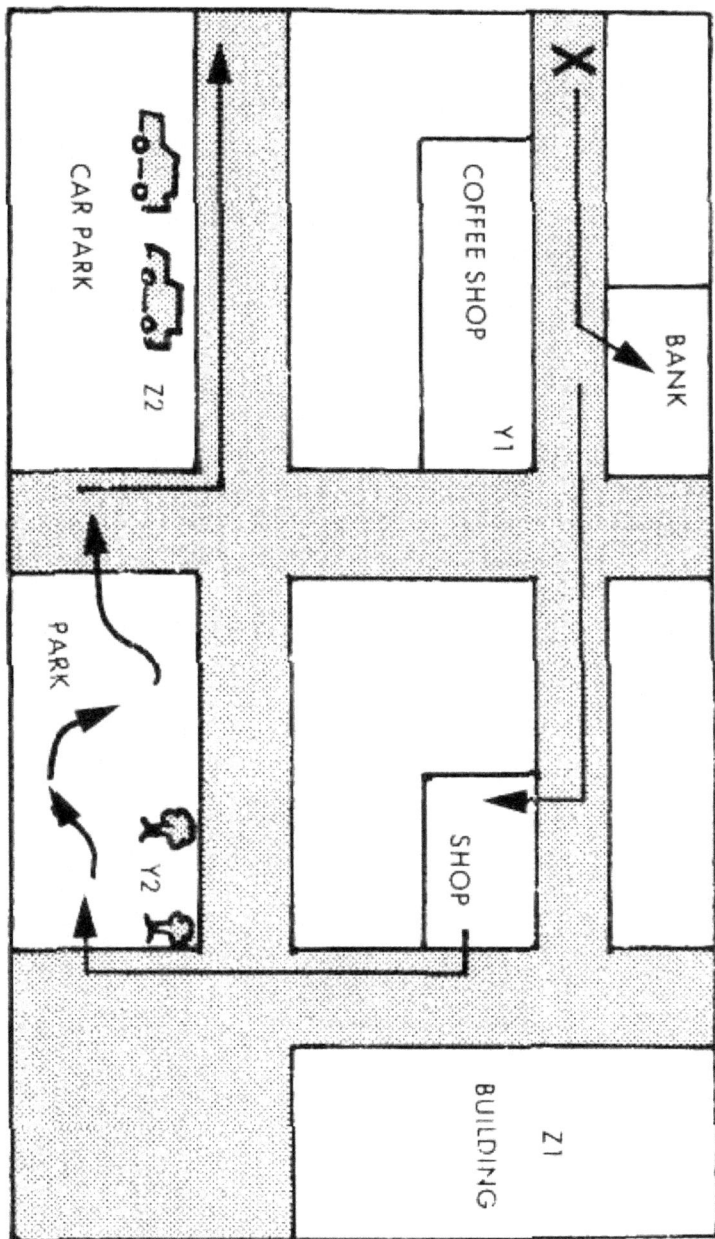

Such check points could be:

- From inside a coffee shop Y gets a good view of X entering the bank opposite
- Z1 Z is in a building (roof garden, balcony or upper floor window) watching X's progress down the street and into the bookshop
- Y2 Y has moved into park and observes X's wanderings from park bench among the trees
- Z2 Z has time to occupy parked cars in car park with good view of all movement.

After the exercise Y and Z meet to compare notes. What suspicious individuals have they observed? Were such people noticed in X's vicinity on more than just one or two occasions? Was their behaviour strange and were they showing unusual interest in X2 going into check what he was up to? Was a vehicle following them in support and were persons from the vehicle taking it in turns to follow X? Such persons are more easily noticed and remembered in quiet rather than busy areas!

Remember: In order to carry out secret work you must know whether you are under surveillance or are clean!

10. CHECKING BY CAR

There are many ways of countering enemy surveillance when using a vehicle. Be extra observant when approaching your parked car and when driving off. This is the most likely point at which tailing may start from your home, work, friends, meeting place. Be on the lookout for strange cars, with at least two passengers (usually males). When driving off be on the lookout for cars pulling off after you or possibly following you from around the corner. Bear in mind that the enemy may have two or three vehicles in the vicinity, linked by radio. They will try to follow you in an interchanging sequence (the so-called A,B,C technique). Cars A, B and C will constantly exchange positions so as to confuse you.

Ruses:

After driving off it is a useful procedure to make a U-turn and drive away in the opposite direction, forcing any surveillance car into a hurried move. As you proceed, notice vehicles behind you your rear-view mirror is your best friend!

Also pay attention to vehicles travelling ahead which may deliberately allow you to overtake them. Cars waiting ahead of you at junctions, stop street and by the roadside must be noted too. You will often find vehicles travelling behind you for quite a distance, particularly on a main road or link road. Avoid becoming nervous and over-reacting. Do not suddenly speed ahead in the hope of losing them.

Remember that the point of counter-surveillance is to determine whether you are being followed or not. Rather travel at normal speed and then slightly reduce speed, giving normal traffic the chance of overtaking you. If the following vehicle also reduces speed, then begin to accelerate slightly. Is that vehicle copying you? If so, turn off the main road and see if it follows. A further turn or two in a quiet suburb or rural area will establish whether you have a tail.

There are many other ruses to determine this:

• Drive completely around a traffic circle as though you have missed your turn-off;
• Turn into a dead-end street as if by mistake;
• Turn into the driveway of a house or building and out again as if in error;
• Abruptly switch traffic lanes and unexpectedly turn left or right without indicating, but be sure there is no traffic cop about!
• Cross at a traffic light just as it turns red, etc.

Such ruses will force a tail into unusual actions to keep up with you but your actions must appear normal.

Check Route

The Check Route we previously described for checking surveillance by foot can obviously be applied to vehicles. Your check route must be well prepared and should include busy and quiet areas. Also include stops at places such as garages and shops where you can carry out some counter-surveillance on foot. You can carry out your routine by yourself or with assistance. In this case comrades are posted at check points along your route and observe whether you are being tailed. It is a good idea to fit your car with side-view mirrors for better observation, including one for your passenger. At all costs avoid looking over your shoulder (a highly suspicious action!)

Enemy Tracking Device

You should often check underneath your car in case the enemy has placed a tracking device ('bumper bleeper') there. It is a small, battery-operated, magnetically attached gadget that emits a direction signal to a tailing vehicle. This enables the vehicle to remain out of your sight. When you stop for some minutes, however, your trackers will be curious about what you are up to. This will force them to look for you. So your check routine should involve stopping in a quiet or remote area. Get out of your car and into a hidden position from where you can observe any follow-up movement. If you have assistance stop your car at a pre-arranged spot. Your comrades should drive past and check whether a tail vehicle has halted just out of sight down the road.

11. CUTTING THE TAIL

The procedure of eluding those who are following you is called 'cutting the tail'. In order to do this effectively you must study the location or areas where this can be done in advance. When you find yourself in a situation where you need to break surveillance, you deliberately lead those who are following you to a favourable spot where 'cutting the tail' can be achieved.

Change of Clothing:

You urgently need to visit an underground contact. For several days your attempts have been frustrated because you have come to realise that you are being closely watched and followed by the police and their agents. You leave work as usual but carry a shopping bag with a change of clothes. After casually wandering around town you enter a cloakroom or such place where you can quickly change clothing without being seen. It should be a place where other people are constantly entering and leaving. You leave within minutes, casually dressed in a T-shirt and sports cap. Your shirt, jacket and tie are in your shopping bag. A bus area makes it easier to slip away unnoticed. A reversible jacket, pair of glasses and cap kept in a pocket are useful aids for a quick change on the move. Women in particular can make a swift change of clothing with ease, slipping on a wig and coat or even a man's hat and jacket over a pair of jeans to confuse the tail!

Jumping on and off a Bus:

You are being tailed but must get to a secret meeting at all costs. You could spend some time loitering around a busy shopping area giving the impression that you are in no hurry to get anywhere. Just as you notice a bus pulling away from a bus stop you run after it and jump aboard. Keeping a good lookout for your pursuers, you could jump off as it slows down at the next stop and disappear around a busy corner.

Crossing a Busy Street:

You need to be quick and alert for this one! You deliberately lead those following you down a busy street with heavy traffic. When you notice a momentary break in the traffic, you could suddenly sprint across the road as though your life depended on it. By the time the tail has managed to find a break in the traffic and cross after you, you could have disappeared in any number of directions!

Take the Last Taxi in the Rank:

Occupy your time in a leisurely way near a taxi rank. You could be window shopping or drinking tea at a cafe. When you notice that there is only one taxi left at the rank, drop everything and sprint over to it. By the time those following you have summoned up their support cars you could have ordered the taxi to stop and slipped away.

Entering and Exiting a Building:

A large, busy department store with many entrances, stairways, lifts and floors is ideal for this one. After entering the building quickly slip out by another exit. Busy hotels, restaurants, recreation centres, railway stations, arcades, shopping centres etc. are all useful locations for this trick.

Ruses when Driving:

It is more difficult to cut a tail when driving than when on foot because a number of vehicles may be following you in parallel streets. Fast and aggressive driving is necessary. Sudden changes of speed and direction, crossing at a traffic light just as it turns red, and a thorough knowledge of lanes, garages and places where a car may be quickly concealed are possible ways in which you may elude the tail.

Get Lost in a Crowd:

It is particularly difficult for the tail to keep up with you in crowded areas. Know the locality, be prepared, be quick-footed and quick-witted! Be ready to take advantage of large concentrations of people. Workers leaving a factory, spectators at a sports fixture, crowds at a market, cinema, railway station or rally offer all the opportunities you need.

Mix this with the above tactics and you will give those trying to tail you the headache and disappointment they so richly deserve.

12. SECRET COMMUNICATIONS

Communications is vital to any form of human activity. When people become involved in secret work they must master secret forms of communication in order to survive detection and succeed in their aims. Without effective secret communication no underground revolutionary movement can function. In fact effective communication is a pillar of underground work. Yet communication between underground activists is their most vulnerable point.

The enemy, his police, informers and agents are intently watching known and suspect activists. They are looking for the links and contact points between such activists which will give them away. It is often at the point when such activists attempt to contact or communicate with one another that they are observed and their would-be secrets are uncovered. The enemy watches, sees who contacts whom, the pounces, rounding up a whole network of activists and their supporters. But there are many methods and techniques of secret work, simple but special forms of communication, available to revolutionaries to overcome this key problem.

This section discusses these, in order to improve and perfect secret forms of communication. These are used worldwide, including by state security organs, so we are giving nothing away to the enemy. Rather we are attempting to arm our people. These methods are designed to outwit the enemy and to assure continuity of work. The qualities required are reliability, discipline, punctuality, continuity and vigilance – which spells out efficiency in communication.

Before proceeding, however, let us illustrate what we are talking about with an example: C – a member of an underground unit – is meant to meet A and B at a secret venue. C is late and the two others have left. C rushes around town trying to find them at their homes, work place, favourite haunts. C tries phoning them and leaves messages. C is particularly anxious because he has urgent information for them. People start wondering why C is in such a panic and why he is so desperate to contact A and B who are two individuals whom they had never before associated with C. When C finally contacts A and B they are angry with him for two reasons. Firstly, that he came late for the appointment. Secondly, that he violated the rules of secrecy by openly trying to contact them. C offers an acceptable reason for his late-coming (he could prove that his car broke down) and argues that he had urgent information for them. He states that they had failed to make alternative arrangement for a situation such as one of them missing a meeting. Hence, he argues, he had no alternative but to search for them.

The above example is familiar to most activists. It creates two problems for the conduct of secret work. It creates the obvious security danger as well as leading to a breakdown in the continuity of work.

What methods are open to such a unit, or between activists?

To answer this we will be studying two main areas of communication. There are personal and non-personal forms of communication. Personal are when two or more persons meet under special conditions of secrecy. There are various forms of personal meetings, such as regular, reserve, emergency, blind, check and accidental. Then there are various non-personal forms of communication designed to reduce the frequency of personal meetings. Amongst these are such methods as using newspaper columns, public phone boxes, the postal system, radios and the method made famous in spy novels and films, the so-called dead-letter-box or DLB, where messages are passed through secret hiding places. Coding, invisible ink and special terms are used to conceal the true or hidden meaning in messages or conversations.

From this we can immediately see a solution to C's failed meeting with A and B. All they needed to arrange was a reserve meeting place in the event of one or more of them failing to turn up at the initial venue. This is usually at a different time and place to the earlier meeting. The other forms of meetings cover all possibilities.

13. PERSONAL MEETINGS

In the previous section we began to discuss the methods members of an underground unit should use when communicating with one another. The most important requirement that must be solved is how to meet secretly and reliably.

Let us suppose that comrade A has the task of organising an underground unit with B and C. In the interests of secrecy they must, as far as possible, avoid visiting one another at home or at wok. (Such links must be kept to a minimum or even totally avoided so that other people do not have the impression that they are closely connected.)

First of all they need to have a regular or main meeting – let's say every two weeks. For this meeting A lays down three conditions. These are: place, time and legend.

Place of Meeting:

This must be easy to find, approach and leave. It must be a safe place to meet, allowing privacy and a feeling of security. It could be a friend's flat, office, picnic place, beauty spot, beach, park, vehicle, quiet cafe, etc. The possibilities are endless. It is essential that the meeting place be changed from time to time. Sometimes, instead of indicating the meeting place, A might instruct B and C to meet him at different contact points on the route to the meeting such as outside a cinema, bus stop etc. This can provide a greater degree of security. But it is best to begin with the most simple arrangements.

Time:

Date and time of the meeting must be clearly memorised. Punctuality is essential. If anyone fails to arrive at the meeting place within the prearranged time the meeting must be cancelled. As a rule the time for waiting must never exceed ten minutes. Under no circumstances must a comrade proceed to the meeting if he or she finds themselves under surveillance.

Legend:

This is an invented but convincing explanation (cover story) as to why A, B and C are always together at the same place at the same time. The legend will depend on the type of people who are meeting. Suppose A and B are black men and C is an older, white woman. Since it would look unusual and attract attention if they met at a park or picnic place, A has decided on an office which C has loaned from a reliable friend. They meet at 5.30pm when the office is empty. C has told her friend that she requires the premises in order to interview some people for a job or some story to that effect. On the desk she will have interview notes and other documents to support her story and B and C will carry job applications or references. If anyone interrupts the meeting or if they are questioned later, they will have a convincing explanation for their meeting.

Order of the Meeting:

At the start of the meeting A checks on the well-being and security of each comrade, particularly whether everything was in order on their route to the meeting. Did they check for possible surveillance? Next A will inform them of the legend for the meeting. Then, before business is discussed,

A will pass around a piece of paper with the time and place of the next meeting written on it. Nothing is spoken in case the meeting is 'bugged'. This matter is settled in case they are interrupted and have to leave the meeting in a hurry. In such an event they already know the conditions for the next meeting and continuity of contact is assured.

Reserve Meeting:

In arranging the regular meeting of the unit, A takes into account the possibility of one or more of them failing to get to that meeting. He therefore explains the conditions for a reserve meeting. These also include place, time and legend. Whilst the time for a reserve meeting may be the same as a regular meeting (but obviously on a different day), the place must always differ. A instructs them that if a regular meeting fails to take place they must automatically meet two days later at such-and-such a time and place. The conditions for a reserve meeting might be kept constant, not changing as often as those of the regular meeting, because the need for such a meeting may not often arise. But A takes care to remind the comrades of these conditions at every regular meeting.

Having arranged conditions for both regular and reserve meetings, A feels confident that he has organised reliability and continuity of such contact. It is necessary for all to observe the rules of secrecy, and to be punctual, reliable, disciplined and vigilant about such meetings.

But what if comrade A needs to see B and C suddenly and urgently and cannot wait for the regular meeting?

14. EMERGENCY AND CHECK MEETINGS

The leader of an underground unit, comrade A, has arranged regular and reserve meetings with B and C. This allows for reliability and continuity of contact in the course of their secret work. This has been progressing well. Comrade A decides to organise other forms of meetings with them because of the complexity of work.

Emergency Meeting:

The comrades have found that they sometimes need to meet urgently between their regular meetings. An emergency meeting is for the rapid establishment of contact should the comrades need to see each other between the set meetings.

There are similar conditions as for a regular meeting such as: Time, Place and Legend. The additional element is a signal for calling the meeting. This signal might be used by either the unit leader A or the other cell members, when they need to convey urgent information. A confirmation signal is also necessary which indicates that the call signal has been seen or understood. This must never be placed at the same location as the call signal.

Signals:

These are prearranged signs, phrases, words, marks or objects put in specified places such as on objects in the streets, on buildings etc., or specified phrases in postcards, letters, on the telephone etc.

Example of Emergency Meeting:

Comrade A has directed that the venue for the unit's Emergency meeting is a certain park bench beside a lake. The time is for 5.30pm on the same day that the call signal is used. As with Regular meetings he also indicates a Reserve venue for the Emergency meeting. Comrade A arranges different call signals for B and C, which they can also use if they need to summon him.

Call and Answer Signal for B:

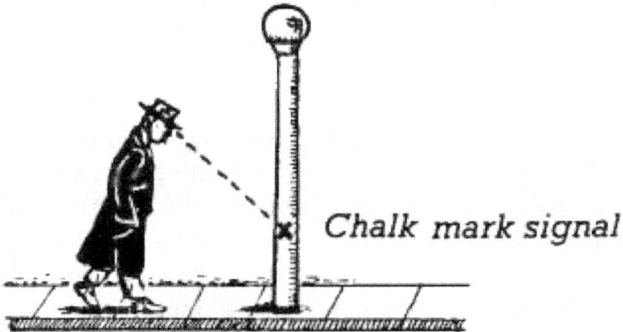

Chalk mark signal

This signal could be a 'chalk mark' placed by A on a certain lamp-post. Comrade A knows that B walks passed the pole every morning at a certain time on his way to work. B must always be on the look-out for the chalk mark. This could simply be the letter 'X' in red chalk. By 2pm. that day B must have responded with the confirmation signal. This could be a piece of coloured string wound round a fence near a bus stop. It could equally be a piece of blue chalk crushed into the pavement by the steps of a building or some graffiti scrawled on a poster (in other words anything clear, visible and innocent-looking). The two comrades can now expect to meet each other at the park bench later that day.

Call and Answer Signal for C:

C has a telephone at home. Before she leaves for work, comrade A phones her from a public call-box. He pretends to dial a wrong number. 'Good morning, is that Express Dairy?' he asks. 'Sorry, wrong number', C replies and adds: 'Not such a good morning, you got me out of the bath'. This is C's innocent way of confirming that she has understood the signal. Obviously such a signal cannot be repeated.

Check Meeting

This is a 'meeting' between the unit leader and a subordinate comrade to establish only through visual contact whether the comrade is all right. Such a check-up becomes necessary when a comrade has been in some form of danger and where direct physical contact is unsafe to attempt, such as if the comrade has been questioned by the police or been under surveillance.

There are a number of conditions for such a meeting: Date and Time; Place or Route of movement; Actions; Legend; Signals – indicating danger or well-being.

Example of Check Meeting:

C has been questioned by the police. As a result contact with her has been cut. After a few days comrade A wants to check how she is and calls her through a signal to a Check meeting.

At 4pm on the day following the call signal C goes shopping. She wears a yellow scarf indicating that she was subject to mild questioning and that everything has appeared normal since. She follows a route which takes her past the Post Office by 4.20pm. She does not know where A is but he has taken up a position which conceals his presence and gives him a good view of C. He is also able to observe whether C is being followed. On passing the Post Office C stops to blow her nose. This is to reinforce her feeling that everything is now normal. It is for A to decide whether to restore contact with C or to leave her on 'ice' for a while longer, subjecting her to further checks.

15. BLIND MEETING

The leader of an underground unit, comrade A, receives instructions from the leadership to meet comrade D. Comrade D is a new recruit, whom the leadership are assigning to A's unit. A and D are strangers to one another. Conditions are therefore drawn up for a Blind Meeting – that is a meeting between two underground workers who are unknown to one another.

Recognition signs and passwords

There are similar conditions as for regular and other forms of meeting, such as date, time, place, action of subordinate and legend. In addition, there is the necessity for recognition signs and passwords, which are to aid in identification.

The recognition signs enable the commander or senior, in this case A, to identify the subordinate from a safe distance and at close quarters. Two recognition signs are therefore needed.

The passwords, including the reply, are specially prepared words and phrases which are exchanged and give the go-ahead for the contact to begin. These signs and phrases must look normal and not attract attention to outsiders.

At this point the reader should prepare an example for a blind meeting and compare it with the example we have given. Our example has been purposely printed upside down to encourage the reader to participate in this suggested exercise. Do remember that all the examples given in our series are also read by the enemy, so do not blindly copy them. They are suggestions to assist activists with their own ideas.

Example of Blind Meeting Place: Toyshop on Smith Street. Date and Time: December 20th, 6pm.

Action: Comrade D to walk down street in easterly direction, to stop at Toyshop and gaze at toy display for five minutes.

Legend: D is simply walking about town carrying out window shopping. When A makes contact they are to behave as though they are strangers who have just struck up a friendship.

Recognition signs: D carries an OK Bazaars shopping bag. The words 'OK' have been underlined with a black pen (for close-up recognition).

Passwords:

A: Pardon me, but do you know whether this shop sells children's books?

B: I don't know. There are only toys in the window.

A: I prefer to give books for presents.

Note: The opening phrase will be used by A after he has observed D's movements and satisfied himself that the recognition signs are correct and that D has not been followed. A completes the passwords with a closing phrase which satisfies D that A is the correct contact. The two can now walk off together or A might suggest a further meeting somewhere else.

Brush Meeting

This is a brief meeting where material is quickly and silently passed from one comrade to another. Conditions for such a meeting, such as place, time and action, are carefully planned beforehand. No conversation takes place. Money, reports or instructions are swiftly transferred. Split-second timing is necessary and contact must take place in a dead zone i.e. in areas where passing the material cannot be seen.

For example, as D walks down the steps of a department store A passes D and drops a small package into D's shopping bag.

'Accidental' Meeting

This is, in fact, a deliberate contact made by the commander which comes as a surprise to the subordinate. In other words, it takes place without the subordinate's foreknowledge.

An 'accidental' meeting takes place where:

1. there has been a breakdown in communication.
2. the subordinate is not fully trusted and the commander wants to have an 'unexpected' talk with him or her.

The commander must have good knowledge of the subordinate's movements and plan his or her actions before, during and after the meeting.

16. NON-PERSONAL COMMUNICATION

Comrade A has been mainly relying on personal forms of communication to run the underground unit. With the police stepping up their search for revolutionary activists he decides to increase the use of non-personal communication.

These are forms of secret communication carried out without direct contact. These do not replace the essential meetings of the unit, but reduce the number of times the comrades need to meet, thereby minimising the risks.

The Main Forms:

These are telephone, postal system, press, signals, radio and dead letter box (DLB). The first three are in everyday use and can be used for secret work if correctly exploited. Signals can be used as part of the other forms or as a system on their own. Radio communication (coded) will be used by higher organs of the Movement and not by a unit like A's. The DLB is the most effective way of passing on material and information without personal contact.

Comrade A introduces these methods cautiously because misunderstandings are possible. People prefer face-to-face contact so confidence and skill must be developed.

Telephone, Post and Press:

These are reliable means of secret communication if used properly. Used carelessly in the past they have been the source of countless arrests. The enemy intercepts telephone calls and mail going to known activists and those they regard as suspicious. Phone calls can be traced and telexes as well as letters intercepted. International communication is especially vulnerable. For example, a phone call from Botswana to Soweto is likely to arouse the enemy's interest. What is required are safe phones and addresses through which can be passed innocent-sounding messages for calling meetings, re-establishing contact, warning of danger, etc.

Telephone:

This allows for the urgent transmission of a signal or message. The telephone must be used with a reliable and convincing coding system and legend. Under no circumstances must the phone be used for involved discussion on sensitive topics.

Comrade A has already used the phone to call C to an emergency meeting (See No 14 of this series). The arrangement was that he pretended to dial a wrong number. This was the signal to meet at a pre-arranged place and time.

Up to now he has been meeting with her to collect propaganda material. He now wishes to signal her when to pick it up herself, but prefers to avoid phoning her at home or work. If she takes lunch regularly at a certain cafe or is at a sports club at a certain time or near a public phone, he knows how to reach her when he wishes.

A simple call such as the following is required: 'Is that Miss So-and-So? This is Ndlovu here. I believe you want to buy my Ford Escort? If so, you can view it tomorrow.' This could mean that C must collect the propaganda material at a certain place in two days time. The reference to a car is a code for picking up propaganda material; Ndlovu is the code name for the pick-up place; tomorrow means two days time (two days time would mean three days).

Post:

This can be used to transmit similar messages as above. A telegram or greeting card with the message that 'Uncle Morris is having an operation' could be a warning from A to C to cut contact and lie low until further notice because of possible danger. The use of a particular kind of picture postcard could be a signal for a meeting at a pre-arranged place ten days after the date on the card. Signals can be contained in the form the sender writes the address, the date or the greeting. 'My dear friend' together with the fictitious address of the sender – 'No 168 Fox Street' – means to be ready for a leaflet distribution and meet at 16 hours on the 8th of the month at a venue code-named 'Fox'.

Many such forms of signals can be used in letters. Even the way the postage stamp is placed can be of significance.

Press:

This is the use of the classified ads section: *'Candy I miss you. Please remember our Anniversary of the 22nd, love Alan'*. This could be A's arrangement for re-establishing contact with C if she has gone into hiding. The venue and time will have been pre-arranged, but the advert will signal the day. Such ads give many possibilities not only in the press but on notice boards in colleges, hostels, shopping centres, and so on.

17. SIGNALS

Comrade A has been introducing various forms of Non-Personal Communications (NPC) to his underground unit. At times he has carefully used the telephone, post and press to pass on innocent-sounding messages, (see No.16 of this series). Key phrases, spoken and written, have acted as signals for calling meetings, warning of danger etc. He has also used graphic signals, such as a chalk mark on a lamp post, or an object like a coloured piece of string tied to a fence, as call and answer signs (see No.14).

Signals can be used for a variety of reasons and are essential in secret work. They greatly improve the level of security of the underground and help to avoid detection by the enemy forces.

Everyday Signals

The everyday use of signals shows how useful they are in conveying messages, and what an endless variety exists. Road traffic is impossible without traffic lights (where colour carries the message) and road signs (where symbols or graphics are used). Consider how hand signals are used in different ways not only to direct traffic but for countless purposes from sport to soldiers on patrol. Everybody uses the thumbs-up signal to show that all is well. Consider how police and robbers use signals and you will realise how important they are for underground work. In fact in introducing this topic to his unit Comrade A asks them to give examples of everyday signals. The reader should test his or her imagination in this respect.

For our purpose signals are divided into TYPE and USAGE.

Type:

Sound – voice, music, whistle, animal sound, knocking etc.
Colour – all the hues of the rainbow!
Graphic – drawing, figures, letters, numbers, marks, graffiti, symbols etc.
Actions – behaviour/movement of a person or vehicle.
Objects – the placing or movement of anything from sticks and stones to flower pots and flags.

Use:

To call all forms of meetings; to instruct people to report to a certain venue or individual; to instruct people to prepare for a certain task or action; to inform of danger or well-being; to indicate that a task has been carried out; to indicate a presence or absence of surveillance; to indicate recognition between people.

Whatever signals are invented to cover the needs of the unit they must be simple, easy to understand and not attract attention.

Here are some examples of how signals can be used: One example is included which is bad from the security point of view. See if you can spot it. Consider each example in terms of type and usage:

• Comrade A draws a red arrow on a wall to call B to an emergency meeting.

• D whistles a warning to C, who is slipping a leaflet under a door, indicating that someone is approaching.

• B stops at a postbox and blows his nose, indicating to A, observing from a safe distance, that he is being followed.

• D hangs only blue washing on his clothes line to indicate that the police have visited him and that he believes he is in danger.

• B enters a hotel wearing a suit with a pink carnation and orders a bottle of champagne. These are signals to C that she should join him for a secret discussion.

• C, having to deliver weapons to 'Esther', whom she has not met before, must park her car at a rest-spot venue on the highway. C places a tissue-box on the dash-board and drinks a can of cola. These are the recognition signals for E to approach her and ask the way to the nearest petrol station. This phrase and a Mickey-Mouse key-ring held by E are the signs which show C that E is her blind contact. (Note: both will use false number plates on their cars to remain anonymous from each other).

• C places a strip of coloured sticky tape inside a public telephone box to inform A that she has successfully delivered weapons to E.

The bad example? D's pink carnation and champagne draws unwanted attention.

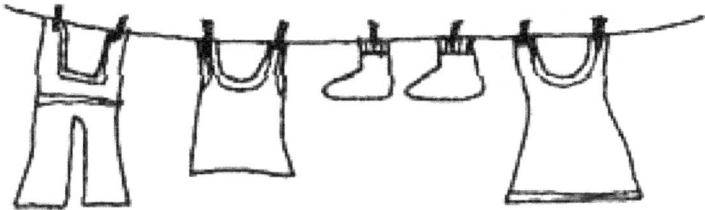

18. DEAD LETTER BOX

Comrade A's underground unit has been mastering forms of Non-Personal Communication to make their work secret and efficient. Comrade A feels they now have sufficient experience to use the DLB, sometimes called a 'dead drop', to pass literature, reports and funds between one another.

The DLB

It is a hiding place such as a hollow in a tree or the place under the floorboards. It is used like a 'post box' to pass material between two people.

To give a definition: A DLB is a natural or man-made hiding place for the storage and transfer of material.

It can be a large space for hiding weapons or small for messages. It can be located inside buildings or out of doors; in town or countryside. It can be in natural spaces such as the tree or floorboards, or manufactured by the operative, such as a hollowed out fence pole or a hole in the ground. It is always camouflaged.

Selecting the DLB

It is very important to carefully select the place where the DLB is to be located. Follow the rules:

● It must be easy to describe and find. Avoid complicated or confusing descriptions which make it difficult for your partner to find it.

● It must be safe and secure. It must be well concealed from casual onlookers. Beware of places where children play, gardeners work or tramps hang-out. It must not be near enemy bases or places where guards are on duty. It must not be overlooked by buildings and windows.

● It must allow for safe deposit and removal of material. The operatives must feel secure about their actions in depositing and removing material. They must be able to check whether they are being watched. The place must be in keeping with their public image and legend.

● It must allow for weather conditions and time of day. DLBs can be exposed or damaged by rain or flooding. Some locations may be suspicious to approach by day and dangerous by night.

Preparation

This involves constructing and camouflaging the DLB; making a diagram; working out a signal system and security arrangements. If you are burying the material put it in a tin, bottle or weather-proof container.

● Once you have selected the place for your DLB you will have to prepare it. This will usually take place under cover of night whether you are digging a hole or hollowing out a cavity in a tree and camouflaging it.

● You will have to make an accurate description, preferably including a simple diagram.

● You will have to work out a signal system for yourself and partner indicating deposit and removal of material.

● Finally, work out a check route to and from the DLB and a legend for being there.

Example of DLB

Comrade A has spotted a loose brick in a wall. The wall is located along a little used path and shielded by trees. At night he hollows-out a space behind the brick, large enough to take a small package. The loose brick is the tenth along the wall, second row down. The brick fits securely into the wall but can be quickly removed with the use of a nail. The operation takes ten seconds and the footsteps of any stranger approaching can be easily heard.

A's Description of the DLB

"Reference No. DLB 3. 'Loose Brick in wall'
Location: Path leading from Fourth Street to Golf Course
Direction: In Fourth Street, just past the 61 Bus Stop, is the path, with red brick wall on the right, wooden fence on the left. Three paces down the path, on the right, just before a tree, is the DLB, in the brick wall.
The DLB: It is a loose brick, with white paint smudge. As you walk down the path from Fourth Street, it is the tenth brick along the wall, second row from top. In the space between this brick and the ninth brick is a hole. Place a nail into this hole to help prise out the brick. The space behind the brick holds a package wrapped in plastic with dimensions: 12x6x3 cm. After removing the package replace brick using blue tack (or other sealing substance) to hold it in place."

Signals:

1. After A deposits material he ties a piece of red string to a fence signalling that the DLB is 'loaded'.

2. After B removes material from the DLB he draws a chalk mark signal on a pole.

Note: Signals must not be in the DLB's vicinity.

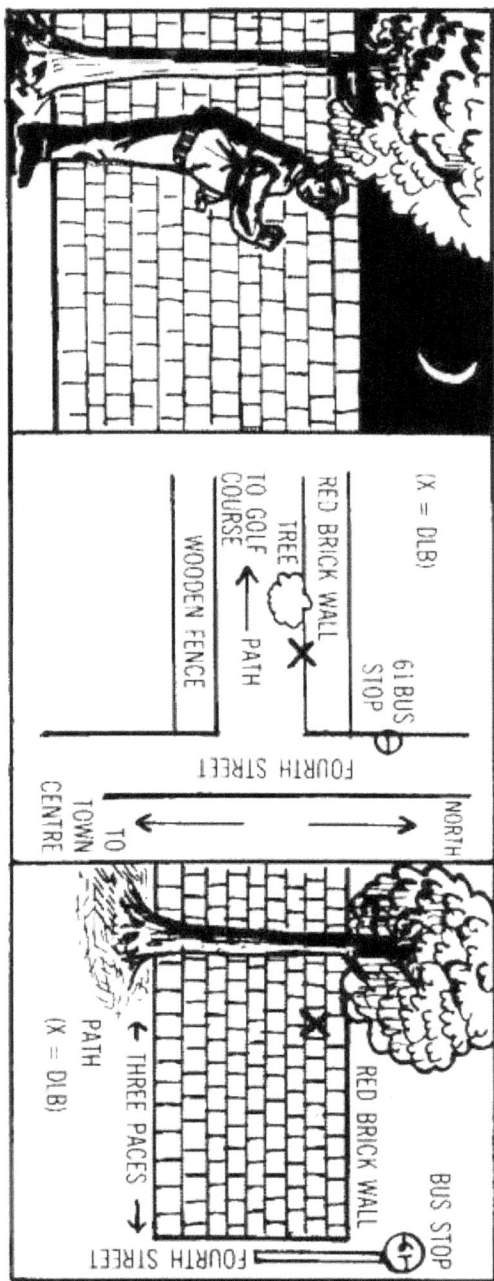

RED BRICK WALL

TREE

TO GOLF COURSE

PATH

WOODEN FENCE

(X = DLB)

61 BUS STOP

FOURTH STREET

TO TOWN CENTRE

NORTH

RED BRICK WALL

PATH

(X = DLB)

THREE PACES

BUS STOP

FOURTH STREET

Carrying Out the Operation

The use of the DLB is an operation which must be carefully planned as follows:

Comrade A:

1. Prepares material (packaging and camouflaging)
2. Checks route for surveillance
3. Observes situation at DLB
4. Places material (if no surveillance)
5. Return route to check for surveillance
6. Places signal indicating deposit
7. Returns home

Comrade B:

1. Sees signal of deposit
2. Checks route
3. Observes situation at DLB
4. Removes material (if no surveillance)
5. Return route to check for surveillance)
6. Places signal of removal
7. Returns home.

Comrade A:

1. Checks signal of removal
2. Removes signals
3. Reports success

Note: It is important that both A and B check that they are not being followed when they go to the DLB and after leaving it.

19. STATIONARY, PORTABLE AND MOBILE DLBs

We have been discussing the use of the dead letter box (DLB) through which underground members secretly pass material to each other. There are various types of DLBs:

1. Stationary DLBs are fixed places such as a camouflaged hole in the ground, hollow tree trunk or fence pole, loose brick in a wall (as described in last issue).

2. Portable DLBs are containers which can be carried and left in innocent places to be picked up, e.g. discarded cigarette pack, hollowed-out stick or fake piece of rock.

3. Mobile DLBs are in different types of transport (car, bus, train, boat or plane) and are used to communicate between operatives who live far apart.

4. Magnetic DLBs: A simple magnet attached to a container increases opportunities for finding places to leave your DLB. With the aid of magnets you are able to clamp your DLB to any metal object such as behind a drain pipe, under the rail of a bridge, under a vehicle, etc. Comrade 'A' will use a variety of DLBs with 'B'. Never use a stationary DLB too often because this increases the risk of being spotted. The advantage of a portable DLB is that the place where it is left can be constantly changed. Because of the danger of a stranger picking it up by chance the time between making the drop and the pick-up by

your partner must not be long.

5. Portable DLB – 'Wooden Stick':
Buy a piece of plastic tubing or pipe. Cut off a 30cm length.
Glue pieces of bark around it to make it look like a twig. With
a little patience you will be surprised at how realistic you can
make it. You have a portable DLB into which you can insert
material. Work out a suitable location where it can be safely
dropped for a pick-up. You can carry it up your sleeve and drop
it in long grass or into a bush near an easy-to-locate reference
point. It must be concealed from passers-by and nosey dogs!
Alternatively you can try hollowing out an actual piece of
branch, or splitting it down the side and gluing it. But you
will probably find the plastic pipe easier to handle and longer-
lasting.

6. Portable DLB – 'Hollow Rock': Experiment with plaster of
Paris (which you can buy from a chemist) and mould it into
the shape of a rock. Allow enough of a hollow to hide material.
With paint and mud you can make it look like a realistic rock.
Carry it to the drop-off point in a shopping bag. (Note: the
above can serve as a portable DLB as well as a useful hiding
place for the storage of sensitive material around the home).

7. Mobile DLB - Comrade 'A' uses the Johannesburg to Durban train to send material to comrades down at the coast. There are numerous hiding places on trains, as with other forms of transport, and if you use magnets the possibilities are increased. Removing a panel in a compartment provides a useful hiding place. Comrade 'A' does this long before the train's departure, before other passengers arrive. He has a telephonic signal system with the Durban comrades to indicate when the material is on its way and how to locate it. They might get on the train before it reaches Durban. Whatever the case, the operational system must be carefully studied at both ends.

20. FAILURE AND HOW TO DEAL WITH IT

Our series would not be complete if we did not deal with failure in the underground and how to react to setbacks.

What do we mean by 'Failure'?

When members of the underground are exposed, arrested or killed, when the underground structure is broken-up or destroyed by the enemy – failure has occurred. Failure can be where PARTIAL only some members are affected or COMPLETE, where the entire network or machinery is smashed. OPEN failures are those that the enemy chooses to publicise. CONCEALED failures occur when the enemy succeeds in infiltrating the underground with its agents but keeps this secret. In this case they make no immediate arrests choosing instead to patiently obtain information over a long period.

Reasons for failure

There are numerous causes of arrests and setbacks.

a) Violating the rules of secrecy:

This is one of the main causes of failure. To carry out secret work successfully everyone must strictly follow the organisational & personal rules of behaviour that have been outlined in this series.

Common violation of the rules are:

• failure to limit the number of links between persons (knowledge of others must be limited)

• not keeping to the principle of vertical lines of communication (eg. a cell leader must not have horizontal contact with other cell leaders but only with a contact from the higher organ)

• failure to compartmentalise or isolate different organs from one another (eg. comrades responsible for producing propaganda must not take part in its distribution)

• poor discipline (eg: loose talk; carelessness with documents; conspicuous or unnatural behaviour etc.)

• poor recruitment practises (eg: failure to check on person's background; failure to test reliability; selecting one's friends without considering genuine qualities etc.)

• failure to use codes and conceal real identities

• weak cover stories

• legends

• poor preparation of operations & meetings

• violating the rule of "knowing only as much as you need to know"

• not using the standard methods of personal and impersonal communications

• inadequate preparation of comrades for arrest and interrogation so that they reveal damaging information.

b) *Weak knowledge of the operational situation:*

This means not paying sufficient attention to the conditions in the area where you carry out your tasks. Comrades are often caught because they failed to study the methods used by the enemy, the time of police patrols, guard system, use of informers etc. Mistakes are made if you fail to take into account the behaviour of local people, cultural mannerisms and habits, forms of dress etc. Knowledge of political, economic, geographic and transport conditions are part of the operational picture.

c) Weakly trained and poorly selected operatives:

The underground can only be as strong as its members. Poorly trained leaders result in weak leadership, weak communication links and poor training of subordinates. This leads to wrong decisions and incorrect behaviour throughout the structure and a whole series of mistakes. Care and caution are the key to the selection of capable leaders and recruitment of operatives.

d) Weak professional, political and personal qualities:

Serious shortcomings in the qualities required for underground work can lead to failure. For example a comrade who is sound politically and has good operational skills but who drinks heavily or gambles can put a machinery at risk. Similarly a person with good professional and personal qualities but who is politically confused can be the cause of failure. And a person with good political understanding and fine personal qualities but who has weak operational capability is best used for legal work.

e) Chance or accident:

An unlucky incident can lead to arrest but is the least likely cause of failure.

Preventing failure

Following the principles and rules of secrecy greatly reduces the possibility of failure – "Prevention is better than cure". But when failure occurs we must already be armed with the plans and procedures for dealing with the situation.

21. DETECTING AND LOCALISING FAILURE

When the principles and rules of secrecy are poorly applied failure and arrests follow. The main dangers come from infiltration by enemy agents or the arrest of comrades on operations. DETECTING failure means to be aware of the danger in good time. LOCALISING failure means to act in order to quickly contain the crisis and prevent the damage spreading. The following are the main points to consider:

1. REVIEW THE MACHINERY:

It is only possible to detect and localise failure if the underground has been built on a solid basis according to the correct organisational principles. A study and review of the structure, lines of communication and the personnel is an essential part of secret work. But it becomes impossible to obtain a clear picture if the underground has been loosely and incorrectly put together and is composed of some unsuitable persons. In such a situation it becomes very difficult to correct mistakes and prevent infiltration. A network which is tightly organised, operates according to the rules of secrecy and is cleared of unsuitable operatives is easier to review and manage.

2. CHECK SUSPECTS:

This is part of the work of reviewing the machinery. It must be carried out discretely so as not to alert the enemy or undermine the confidence of operatives.

1. Review the suspects behaviour, movement and performance;
2. check with co-workers, friends, family;
3. carry out surveillance by the security organ after exhausting the other checks to determine whether there are links with the police.

3. SOME TACTICS OF ENEMY AGENTS:

● they try to win your confidence by smooth talk and compliments;
● they try to arouse your interest by big talk and promises;
● try to get information and names from you which is no business of theirs;
● try to get you to rearrange lines of communication and contact points to help police surveillance;
● may show signs of nervousness, behave oddly, show excessive curiosity;
● may pressurise you to speed up their recruitment or someone they have recommended;
● ignore instructions, fail to observe rules of secrecy;

Note: good comrades can be guilty of lapses in behaviour from time to time, and agents can be very clever. So do not jump to conclusions but study the suspect's behaviour with care and patience. Sooner or later they will make a mistake.

4. LOCALISING FAILURE:

This involves two things: acting against infiltration when it is detected and acting against exposure of the machinery and preventing further arrests, capture of documents, material etc.

a) Acting against infiltration:

The severity of action will depend on the stage reached and the danger posed. The enemy agent may be:

- cut-off without explanation;
- politely cut-off with a good, believable pretext (eg. told the underground unit is being dissolved);
- "frozen" – told they are not being involved because they are being held in reserve;
- arrested and taken out of the country as a prisoner;
- eliminated – where they pose serious danger to the survival of comrades and there is no other way.

b) Avoiding arrest:

The moment it is known that a comrade has been arrested those whose identities he or she could reveal must immediately go into hiding. Most arrests take place because this rule is ignored. Even if it is believed that the arrested comrade is unlikely to break this precaution must be observed.

Everyone must have an "ESCAPE PLAN". This includes an early warning system; assistance; safe hiding place; funds; transport; disguise; new documents of identity;

Endangered comrades may "lie low" until the threat passes or work in another part of the country or leave the country;

All links must be cut with a comrade who has come under enemy suspicion or surveillance. In this case the comrade may be "put on ice" until the danger has passed.

All documents, incriminating material etc. must be destroyed or removed from storage places known to the arrested comrade including from his or her house and place of work;

All comrades must be instructed on how to behave if arrested. They must refuse to give away their fellow comrades and strive to resist even under torture. The longer they resist the more time they give their comrades to disappear and get rid of evidence.

Everything must be done to help the arrested comrade by providing legal representation, publicity, food and reading material if possible, solidarity with the family, organising protest.

END